CO2

& NET ZERO

A Matter of Life or Death

First published in the United Kingdom in 2024

By Kahboom™

©Kahboom Ltd

Printed by KOPA in Lithuania

CO2
and
NET ZERO

Dr Philip Blakeley BA PhD

This book is dedicated to my granddaughter, Morgan, who encouraged me to write this book.

Dr Philip Blakeley has a Bachelors degree in Physics and Chemistry and a PhD in Physics from the University of Keele in the UK.

He has worked as a freelance physicist assisting companies to use scientific principles to develop new products. Examples are the development of the sea wave turbine which has successfully generated electricity on the coast of Islay in Scotland, and designing sub-sea equipment to capture and analyse methane concentrations at 3km deep off shore of San Francisco.

Dr William Happer is the Cyrus Fogg Bracket Professor of Physics emeritus at Princeton University.

He studied physics at the University of North Carolina, graduating in 1960. He earned his doctorate at Princeton University in 1964. He has published significant papers in several subjects in physics and the sciences. More recently he has concentrated on the physics related to global warming.

He was appointed to the U.S. Department of Energy Office of Science by President G W Bush and was later appointed by President Trump to the U. S. National Security Council to report on the relationship between Carbon Dioxide emissions and global warming.

Introduction by Professor William Happer

Policies to "save the planet" from supposed threats of greenhouse gases, most notably carbon dioxide (CO_2) are beginning to have substantial costs. Eliminating net emissions of CO_2 from the combustion of fossil fuels, say by the year 2050, will be enormously expensive: greater than the costs of world wars. It will also mean giving up what were once considered inalienable rights to "life, liberty and the pursuit of happiness," rights won at great cost by earlier generations. Those rights are inconsistent with a brave new world of net zero CO_2 emissions.

If there really were a climate emergency, perhaps the sacrifices of "net zero" would be worthwhile. But there is no climate emergency and there will not be one. The sacrifice will be all pain with no gain. If the entire world were to achieve net zero by the year 2050, only about 0.3 C of warming would be averted, according to "consensus" estimates of the United Nations, and less than 0.1 C for more reasonable assumptions.

Most people are so busy trying to live worthwhile lives, that they don't have time to learn how Earth's climate works, or to judge for themselves whether it is really under threat. But given the sacrifices we, our children and grandchildren are being asked to make, we should all learn more. Phillip Blakely's book, *CO$_2$ and Net Zero,* provides an efficient and interesting way to master many basic facts.

Earth's climate involves lots of hard sciences: physics, chemistry, oceanography, solar physics, etc. This book, *CO$_2$ and Net Zero,* provides a very thorough introduction to the basic science of climate, without any of the complicated equations used by more academic books. But there are many interesting figures and graphs which illustrate key messages.

Of particular interest is the stress on the geological past, when atmospheric CO$_2$ concentrations have usually been higher than those today, and almost never lower. Earth has already experimented with much more atmospheric CO$_2$ than today. The geological record shows that the biosphere loves more CO$_2$. Deciphering parts of the geological record requires conjectures, some more defensible than others. But even with that caveat, there is little geological support for the dogma that CO$_2$ is the "control knob" of climate. Climate is controlled by many factors, not just by greenhouse gases. Furthermore, the most important greenhouse gas is water vapor, not CO$_2$. Clouds add to water's dominant role in Earth's climate.

Future generations will look back on the current demonization of CO$_2$, with the same bewilderment that we look back on medieval witch hunts to stop bad weather. *CO$_2$ and Net Zero* clearly explains that CO$_2$ is not a ``pollutant." People, animals, and plants are made of carbon-based molecules. Along with water and sunlight CO_2 is the basis for almost all life on Earth.

This book is written in clear and simple English. It will help those who have never had time to devote to climate issues become well informed. Those who already know a lot will gain a deeper understanding of this fascinating and important topic. One of the most important lessons of the book is how to discuss complicated scientific issues clearly with intelligent people who have little or no prior familiarity with a complicated scientific issue like Earth's climate.

A rare occurrence of liquid Carbon Dioxide venting from a sub-sea volcano 1600m deep in the Mariana Trench located in the Pacific Ocean.

CO2 and NET ZERO

We are living in an age when science and technology, developed in the last three centuries, has enabled us to examine the condition of the Earth in a way that had not been possible previously.

We can measure the temperature of our entire planet accurately from satellites on a daily basis: we can monitor the weather globally: the sea currents, wind speeds, sea water levels, extent of ice cover, earthquakes, forest fires, air pressure, the concentration of the gases that make up the atmosphere, and dust levels - all are measured continuously.

In order to investigate these same parameters throughout the history of our developing planet, we can get a good idea by examining tree rings, fossils, geological features and gases trapped in ice cores extracted from great depths.

The data from these measurements shows that the Earth's climate has always been changing - not just from day to day or year to year, but over long periods of time going back to the time before any life evolved. And one feature of the climate change in recent years is that the planet is getting warmer. This is of concern, because if the temperature continues to increase, then it will have significant effects on all our lives and the lives of the animals and plants that inhabit the land and sea.

So why is this happening? There are many factors that are known to affect the Earth's temperature. But one that stands out is that the Carbon Dioxide concentration in the atmosphere has been steadily increasing during the same period and it is known to trap heat from the Sun. So is this the cause, or part of the cause? Since we humans produce CO_2, are we the main culprits? And if this is the case, then is there anything that we can do about it?

In response to this data, some governments have set out to reduce the Carbon Dioxide we humans produce by reducing the amount of fossil fuels we use. The aim is to cut the net amount of CO_2 we produce to net zero.

The assumption is that this will cause the Earth's average temperature to fall by 1.5C to an average of 13.5C. This would cause the ice cover at the North and South Poles to increase, would slightly lower the sea level and would make our environment that much cooler.

Reducing the Carbon Dioxide concentration by the amount predicted to achieve this cooling effect would also reduce the area where plants and trees grow and would reduce the yield of crops that we require to eat.

It would mean that we need to change how we heat our houses and would usually result in our living space being colder. Since a lot of the CO_2 we produce comes from the fuel needed to travel, we would need to change to electric vehicles which are recharged from Solar or Wind energy.

This will require all the countries on Earth to enact the same policies. And will require major changes in the way of life for the entire population of our planet.

However, the net zero plan makes the following assumptions:

1. It assumes that the temperature rise is predominantly caused by the increased amount of CO_2.

2. It assumes that the measures taken will reduce the concentration of CO_2 in the atmosphere.

3. It assumes that by reducing our production of CO_2, it will actually reduce the Earth's temperature.

We shall investigate these assumptions in the following pages.

Global Warming

Planet Earth is getting warmer. The average global temperature has risen by 0.9°C (1.6°F) since 1940 until 2024.

The data from NASA regarding this effect is shown in this chart.

Average temperature change of the Earth 1880 to 2020 in °C

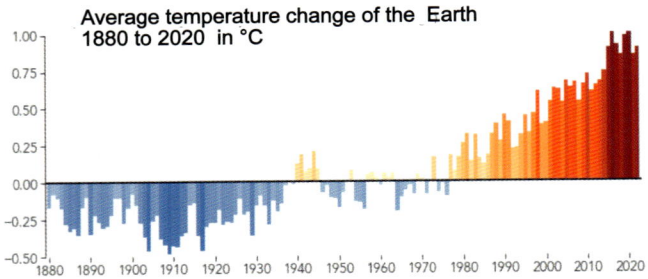

During this time, the concentration of CO_2 in the atmosphere has increased from 0.029% to 0.042%. (290 ppm to 420 parts per million). This is shown in the chart below which is based on data in 2024 provided by the European Environment Agency.

Parts per million CO_2

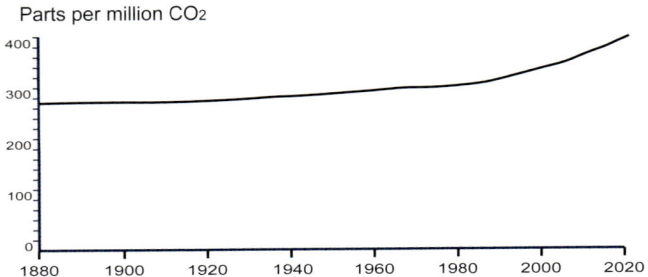

The currently accepted view is that this increase in CO_2 is causing the temperature increase and this book examines this claim, and uses reports from scientists and research institutes to consider to what extent CO_2 is the main culprit for global warming.

We already see a clue that the the cause may be more complex than is currently claimed just by comparing the two charts on the previous page. We see that CO_2 concentrations were slowly increasing from 1880, but for a period of 100 years, the Earth was actually cooling. And in the second half of the 20th century there was concern that the Earth would experience a new ice age.

River Thames in London frozen during the winter of 1963
Credit PA

In the unusually cold period at the end of the 20th century there was even a proposal to spray fine carbon dust over the ice caps to cause melting so that they would not continue to grow.

However, in order to understand what is causing the currently observed global warming, we need to look at much longer time frames than this recent cold event. By cherry picking particular periods it is easy to come up with an incorrect conclusion about the role of CO_2. We need to have a much broader view from long time periods.

The growing size of the human population is producing year-on-year increasing quantities of this gas so it is vital to know if we are turning our planet into an increasingly hotter home which will bring about the extinction of many animal and plant species both on land and in the sea. We need to know if the human CO_2 emissions are contributing to global warming. And to be sure about this we need to identify any other natural processes which could also be causing the increase in temperature that we have been experiencing in recent times.

The current response of governments around the world is to spend enormous sums of money trying to reduce the CO_2 emissions in pursuit of 'Net Zero' - The target is to remove or prevent any human activities that add any additional Carbon Dioxide at all to the existing concentration.

The World Economic Forum commissioned a report in 2022 from the consultancy firm McKinsey, which says '*total global spending by governments, businesses and individuals on energy and land-use systems will need to rise by $3.5 trillion a year, every year, if we are to have any chance of getting to net-zero in 2050.*' That's a 60% increase on today's level of investment and is equivalent to half of global corporate profits, a quarter of world tax revenue and 7% of household spending.

As a result, governments in many countries - mostly the western nations - have already implemented plans to reduce Carbon emissions produced by human activities. The message to the populations of these countries is that by reducing the amount of CO_2 we are emitting, the average temperature of the Earth will cease to increase and then fall to that which was experienced in the 1800s.

This is why it is very important to be sure that CO_2 is the main cause of global warming, and that the net-zero plan will actually stop any further temperature increase. However, there are several other natural processes taking place which may well dominate Earth's temperature which would mean that simply controlling CO_2 would have little effect.

In the following pages we will use scientific reports to look at how CO_2 has influenced the Earth's temperature and also look at other effects which are currently being overlooked.

This picture shows smoke and dust in the atmosphere.
Carbon Dioxide is completely colorless.

What is Carbon Dioxide?

Carbon Dioxide is a colorless, odourless gas. It exists as molecules where one atom of Carbon is bonded to two atoms of Oxygen which is often written as CO_2.

$$O=C=O$$

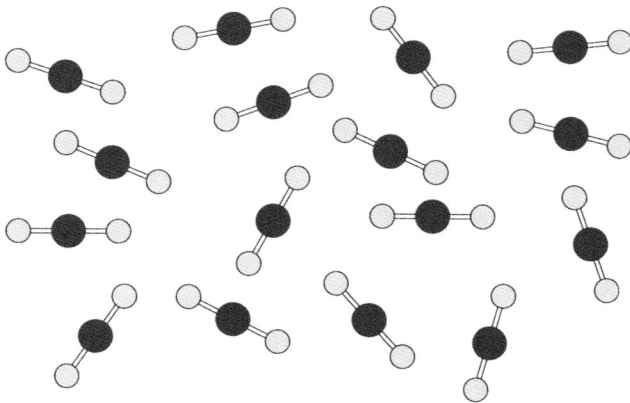

Whenever you look out and see misty, smokey, or dusty air. That is not Carbon Dioxide. CO_2 really is invisible to our eyes.

When we breath air outside, this includes a tiny amount of Carbon Dioxide making up 420 parts per million. Our bodies do not need this so we breath it out again, together with a very small extra amount, which is a result of us digesting our food.

Composition of the Earth's Atmosphere

Nitrogen	78.08%
Oxygen	20.95%
Argon	0.92%
Carbon Dioxide	0.04%

+ very small traces of other gases to make up 100%

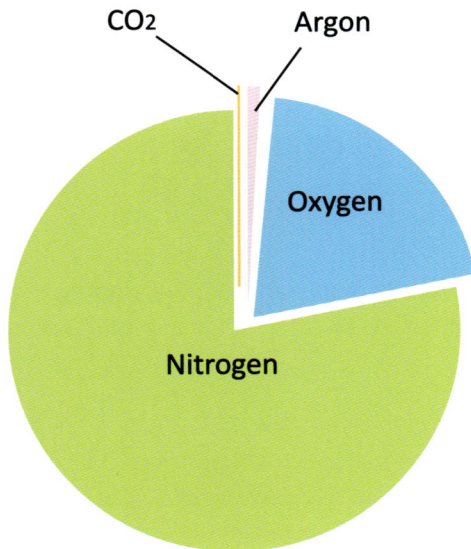

If we are in a room with other people, then the air will contain typically 1500 ppm, and we have no problem with this. The recognised safety limit for CO_2 concentration is 5000 ppm for 8 hours which would only occur if you are in an enclosed space for a long period of time.

The figure on page 20 shows the composition of the Earth's atmosphere at the present time. It forms a thin layer about 10 km (6 miles) thick and in this relatively small zone, together with the oceans, all of life co-exists. This atmospheric layer, together with the land masses and oceans is called the biosphere which is the home for all animals. plants, fungi, and bacteria.

Although CO_2 only makes up a very small part of the air, it is absolutely essential for life to exist. Without it, the Earth would be a dead planet - no plants, and no animals. Evolution, as we know it, could not have occurred. It is not only a key component in the metabolism of living matter, but it also plays an important role in stabilising the Earth's temperature.

The evolution and continued existence of life also depends upon the biosphere maintaining a temperature range of -20 to +40C (0 to 100F). Some species can exist outside these values, but we need to keep in this range on the land surface for life to continue as we know it.

Of course the temperature is higher in the day, cooler at night, warmer in summer and colder in winter, warmer

The Biosphere includes all living things and is 100% dependant on CO$_2$

nearer the equator and colder near the North and South Poles. So in order to check if the temperature of the Earth is changing as the years pass by, it is useful to refer to the average surface temperature. This average is currently 15C, (59F).

The average temperature of the Earth hasn't always been the same as it is now. From its formation about 5 billion years ago, until around 3.5 billion years ago it was too hot for life to form and although it cannot be dated precisely, there is evidence that bacteria-like organisms existed from about 3.5 billion years ago.

There was then a long period when life was evolving and it was about 550 million years ago, in the Phanerozoic period, that creatures with hard body parts left us a fossil record that we find today. The discovery and analysis of these fossils enabled the temperature and the CO_2 concentration of the atmosphere to be determined with reasonable accuracy. Using that data from the time of the earliest recorded fossils until the present day, we see that the temperature of the Earth has varied a lot. And throughout this period, the composition of the atmosphere has also been changing.

Therefore we conclude the climate on Earth is forever changing and in the following pages we shall see how this has had a controlling influence on the evolution of life. We cannot, therefore, expect the climate to now remain constant.

Armed with the information about temperature and CO_2 levels we can come to an understanding of the historical changes in climate, which gives us an idea of the significance of trends we see in the present day weather patterns.

Where does Carbon Dioxide come from?

We know we live on a planet with Carbon Dioxide in the atmosphere, but where does it come from? First we need to know how the Earth obtained Carbon in its makeup.

The planets were forming about 5 billion years ago from dust particles that had been ejected by the young Sun. These particles were made up of many elements such as Iron, Aluminium, Gold, etc. and amongst this mix was Carbon. The dust particles began to clump together due to their gravitational attraction which ultimately led to the existence of the planets But in the early stages, there were many more smaller planets that collided with each other to form bigger and bigger spheres: Early planets.

As time went by, the larger planets, which were heavier, attracted the smaller ones. One of the results caused by the increasing mass of the larger early planets is that their central cores got hotter and hotter due to the gravitational attraction pulling the material together. The Earth got so hot that many of the elements including Carbon and Oxygen evaporated and were lost into space. At this stage only the heavier, less volatile elements were retained on Earth and there was almost no atmosphere.

The smaller early planets did not heat up in the same way and so retained their Carbon and Oxygen components and

Small cool planets collided with Earth about 4.4 - 3.4 billion years ago.

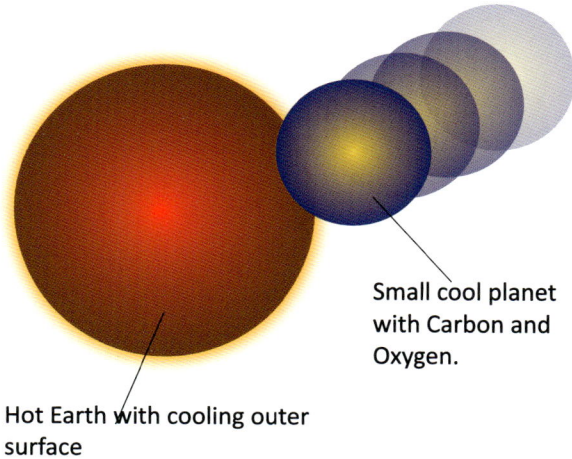

Small cool planet with Carbon and Oxygen.

Hot Earth with cooling outer surface

Reference: Massive impact-induced release of Carbon .....'
Earth and Planetary Science Letters S Marchi et al. June 2016

when many of these collided with the early Earth, they broke up scattering the debris across the upper surface of our planet. These upper layers were cooler and so the more volatile elements including carbon were retained and make up the materials we find today.

Thus the Earth became a planet which contained all the elements which would eventually provide the essential components for all living things, and the Earth was at an ideal temperature so that life could evolve.

As a result of these collisions, there is now 1,850,000,000,000 tons of Carbon on the Earth. Only a very small fraction of this goes to make up CO_2. More than 99% of it is combined with other elements to make up rocks and the shells of sea creatures. We will see next how Carbon is transferred between being in CO_2 as a gas and getting bound up in animals, plants, sea water and rocks. The amount of Carbon on Earth never changes, but the amount that goes to make up CO_2 can change since it is easily taken up and released by biological and geological processes.

Note that that large masses are often often quoted in gigatons.

$$1 \text{ gigaton} = 1 \text{ billion tons}$$

And 1 billion tons is 1,000,000,000 tons.

The Carbon Cycle

The amount of Carbon on Earth has remained unchanged for the past 500 million years during which time all the living plants and animals evolved. But during this period of evolution, the amount of Carbon Dioxide in the atmosphere has not remained constant, Sometimes there has been less CO_2, but for longer periods there has been much more than we have at present. So how does the concentration change?

The diagram on the pages 30 and 31 shows how this happens. Some of the processes on Earth cause Carbon to produce more CO_2, and other processes remove it from the atmosphere.Carbon naturally flows in and out of the land, ocean and through living things as part of the carbon cycle.

Sources of Carbon Dioxide.

All animals – including humans – breathe out CO_2 in the process of respiration. When plants and animals die, their stored carbon is also released as CO_2. Natural processes such as respiration and decay, forest fires and volcanic eruptions add an additional 190.2 billion tonnes of CO_2 to the atmosphere per year.

Then burning of fossil fuels provides an additional 9.1 billion tons per year. The net result is that the burning of fossil fuels by humans makes up 4% of the Carbon Dioxide that is produced every year.

Sinks of Carbon Dioxide (How is it removed?)

Land based plants and phytoplankton in the oceans take in CO_2 in a process called photosynthesis. Soils also store, cycle and emit Carbon Dioxide as part of the carbon cycle. About 30% of the excess CO_2 added to the atmosphere is absorbed by oceans. In total, this results in 190.2 billion tons of CO_2, being removed from the air each year.

Fungi obtain CO_2 not through the toadstools and mushrooms that we see, but via there extensive root systems which extend across most of the planet where they attach into about 90% of all plant species and obtain the CO_2 from the nutrients they extract. This symbiotic relationship also benefits the plants since fungi clean up the dying and dead plants enabling new plants to flourish.

Once the CO_2 is obtained by the fungui, it effectively locks it into the soil. Research carried out by Professor Katie Field at the University of Sheffield in the UK has reported in 2023 that mycorrhizal fungi remove 13.1 billion tons of CO_2 from the atmosphere every year. This is equivalent to 33% of all human emissions.

The ocean holds 60 times more carbon than the atmosphere and absorbs almost 30% of carbon dioxide (CO_2) emissions from human activities. The higher the percentage of CO_2 in the atmosphere the more is extracted by the sea. It is absorbed at the interface between the air and the sea surface and then is taken by microscopic marine life known as zooplankton and when they die they sink to the ocean depths taking the CO_2 with them. The report by Elizabeth Shandwick states that this effect is most likely underestimated.*[Elixabeth Shandwick et al, Australia National Science Agency, 2023}*

Through the process of photosynthesis, plants assimilate carbon dioxide and return some of it to the atmosphere through respiration. The carbon that remains as plant tissue is then consumed by animals or added to the soil as litter when plants die and decompose. A portion of the carbon dioxide which is removed in this way, is estimated to be retained for thousands of years. *[Ecological Society of America]*

These processes are referred to as the Carbon Cycle which is shown on the next page.

In the

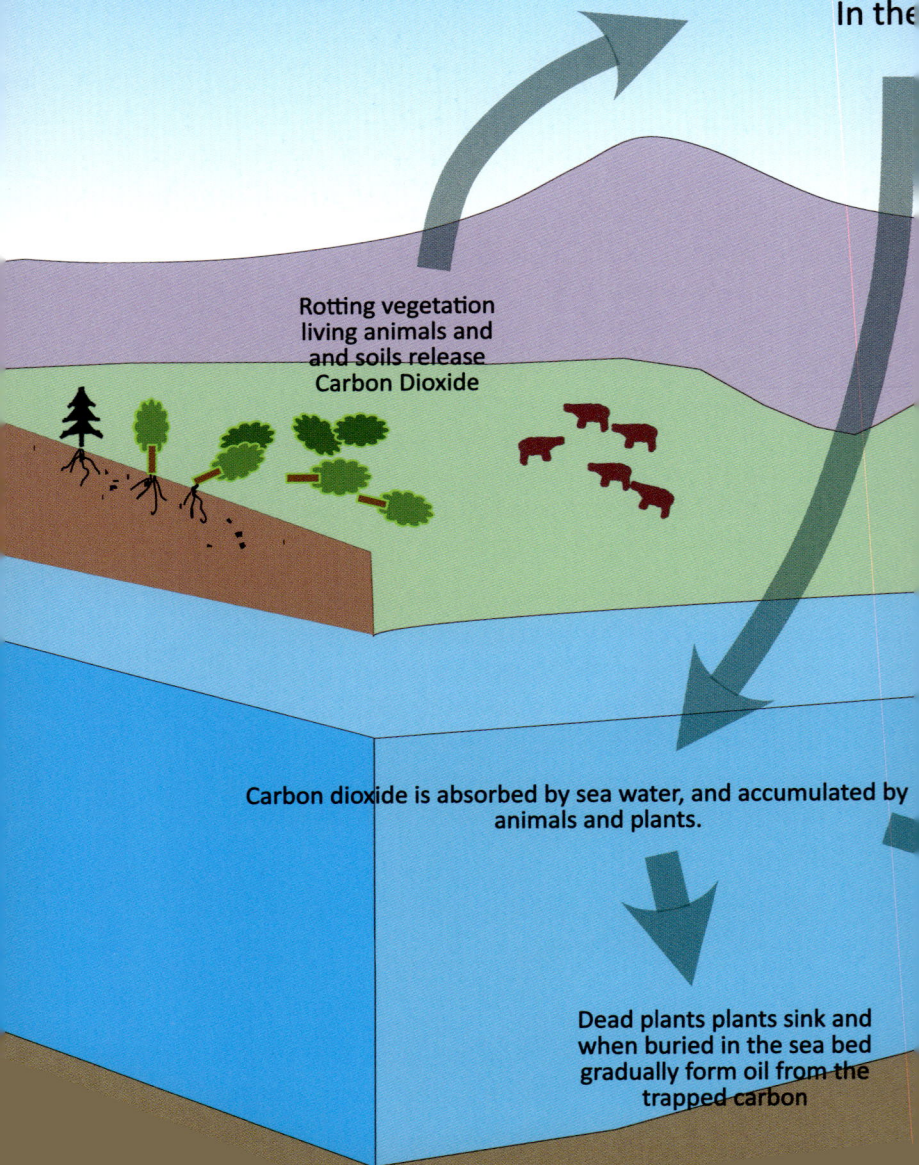

Rotting vegetation
living animals and
and soils release
Carbon Dioxide

Carbon dioxide is absorbed by sea water, and accumulated by
animals and plants.

Dead plants plants sink and
when buried in the sea bed
gradually form oil from the
trapped carbon

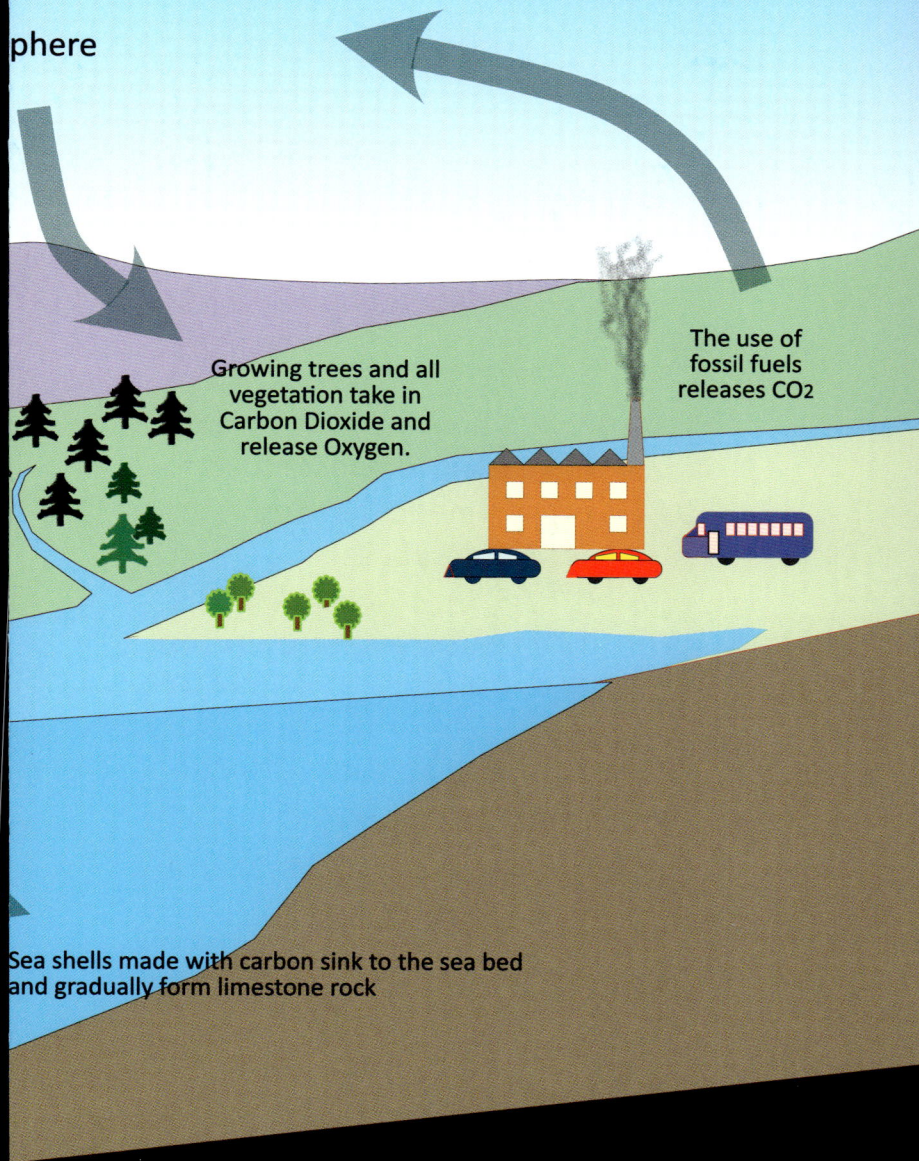

Cycle

phere

Growing trees and all vegetation take in Carbon Dioxide and release Oxygen.

The use of fossil fuels releases CO_2

Sea shells made with carbon sink to the sea bed and gradually form limestone rock

The Earths atmosphere contains 420ppm Carbon Dioxide which is 3284 gigatons.

(ref oakridge national laboratory, 18 July 2020)

Annual Carbon Dioxide emissions due to Human activities currently amount to 29 gigatons

(Ref. European Centre for Medium-Range Weather Forecasts, Nov 2017)

The current additional annual emission of Carbon Dioxide by Humans is therefore 0.88% of the amount already in the atmosphere.

To put this into perspective, laboratory tests have shown that the Termite population on Earth is emitting

50 gigatons of Carbon Dioxide

0.2 gigatons of Hydrogen

And 0.15 gigatons of Methane

Every Year.

Science New Series, Vol. 218, No. 4572 (Nov. 5, 1982), pp. 563-565
Published By: American Association the Advancement of Science

CO2 Contributions from countries that produce more than 1% from human activities.

China	32.9%
United Sates of America	12.6%
India	7.0%
Russia	5.0%
Japan	2.8%
Indonesia	1.8%
Iran	1.8%
Germany	1.8%
South Korea	1.7%
Saudi Arabia	1.6%
Canada	1.5%
Turkey	1.3%
Mexico	1.3%
Brazil	1.2%
South Africa	1.1%
Australia	1.0%
United Kingdom	0.9%

Cassava plants growing in the sunshine of Thailand. They benefit from the higher CO2 levels and are producing increased yields.

What Does CO_2 Do?

CO_2 enables plants to grow and animals to live.

All plants grow by taking in CO_2 and by using the energy of sunlight, along with water and nutrients from the soil, they produce the cells that make up the plant. All plants, trees, mosses, and lichens, all need CO_2. If there was no Carbon Dioxide, the Earth would be a dead planet because in order for animals to evolve, the plants have to come first.

The earliest life on Earth began as microscopic jelly like specks in the sea some 3.5 billion years ago. These living organisms remained small and simple for another 2 billion years . Then around 1 billion years ago, the conditions on Earth changed: it became cooler and the atmosphere included Oxygen and Carbon Dioxide which led to the evolution of plants which were able to grow using energy from the sun. Between 1 billion and 500 million years ago, the plants in the sea and on land evolved, diversified, and became larger. It wasn't until 540 million years ago that animals first appeared.

When the Earth was covered with much more vegetation than we have today, there was more than 1000 parts per million (ppm) Carbon Dioxide in the atmosphere. This number has varied up and down during the last 500 million years. At the present time, there is only 420 ppm CO_2 in the atmosphere and this is part of the reason why there has been a reduction in the plant life and an increase in the spread of deserts.

In fact the concentration of CO_2 has rarely been as low as it is at present and we do not now see the extensive vegetation that previously existed. These days, if farmers want to increase the yield of their crops in greenhouses, they have to increase the CO_2 level back to 1000 ppm.

New research shows that plants happily absorb much of the additional CO_2 that results from human activities.

(Jurgan Knauer et al. Science Advances 2023)

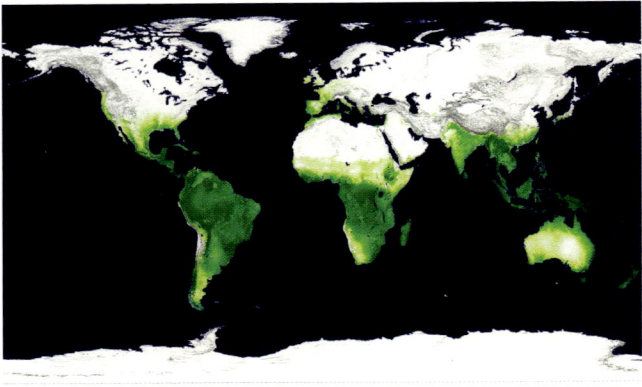

Picture Courtesy of NASA

We already know that plants absorb a huge amount of Carbon Dioxide and by monitoring it's concentration throughout the year we see that during the summer period it decreases significantly due to plant growth. In this illustration we see the result of satellites monitoring CO_2 across the Earth and where plants are extracting large amounts of CO_2 then it is displayed in green. This particular view shows plant activity when it is summer in the southern hemisphere

CO2 helps to stabilise the temperature of the Earth.

Carbon Dioxide in the atmosphere, along with Methane and water vapour help to store the heat energy from the sun. These gases are therefore called greenhouse gases. When the sunlight arrives on the surface of the Earth it warms up the ground. This warm surface would radiate the heat away from the Earth which would cool down quickly at night. But the greenhouse gases retain this heat and slow down the heat loss.

In the last 100 years, the atmospheric CO_2 level has increased from 300 to 420 ppm and there is some concern that this increased greenhouse effect is causing the Earth to heat up and cause significant changes in the climate. However, we shall see next that the amount of Carbon Dioxide in the atmosphere has no direct correlation with the temperature of the Earth. All that the level of CO_2 does is to help stabilise the temperature and enable plants to grow.

The Greenhouse Effect

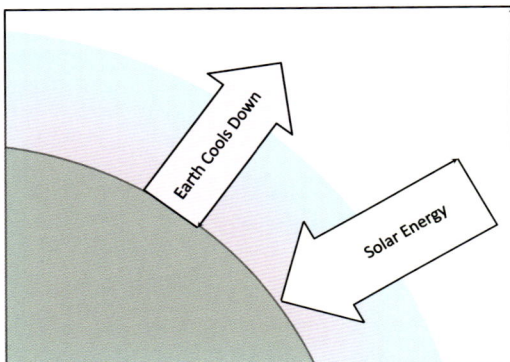

In the day time, the sun heats up the Earth. Some heat escapes the atmosphere and some is retained by the greenhouse gases.

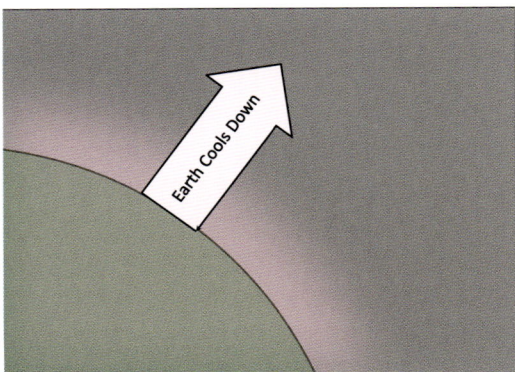

In the night, the Earth cools down but this cooling is slowed down by the greenhouse gases

Gases that cause the Greenhouse effect.

Carbon Dioxide isn't the only gas that results in the greenhouse effect. The main contributors in the atmosphere are:

1. Water vapour, in clear skies.   66.9%
2. Carbon Dioxide   24.6%
3. All other gases   8.5%

By far the greatest contributor to the greenhouse effect is H20 as water vapour and as clouds. Note that clouds do act as a blanket which traps heat in the lower atmosphere, but in the daytime they actually reduce the suns heating effect by reflecting sunlight away from the planet.

[Schmidt et al, J of Geophysical Research, V115, 2010}

If the Earth did not have these greenhouse gases then the average temperature of the planet would be -18°C (0°F) instead of the current temperature of +15°C (+59°F). At the lower temperature of -18°C (0°F) then our planet would be totally covered in ice.

[Karl et al, Science V302, 2003.]

Effects and events that reduce the Carbon Dioxide Concentration.

The Earth itself provides ways of utilising and reducing increases in CO2 without needing any help from human beings.

1. An increase in the extent of plant life.

Increases in CO_2 concentration causes an increase in vegetation. The gas is essential for plant life and in order to grow, plants need CO_2. Satellites continually record the extent of the green vegetation on the planet and the results show that during the recent period when carbon dioxide concentration has increased by 14%, area of plant cover has increased by 11% mainly in the drier arid regions.

[Randell Donohue, Commonwealth Scientific & Industrial Research Organisation, May 2013]

This increase in green territory is because the new plants are taking carbon dioxide from the atmosphere and through the process of photosynthesis, using energy from the sun and nutrients from the soil, enabling them to grow over extended regions. In particular the higher CO_2 levels favour the growth of trees with deep roots which in turn increases the fertility of the soil. Whilst the area of green vegetation is expanding, this is naturally extracting carbon dioxide from the atmosphere.

In addition, the plants that respond best to increased CO_2 levels are found to retain moisture to a greater extent enabling them to flourish in more arid regions.

Plants breath air through small openings in their leaves called stomata. 0.04% of this air is CO_2 which they need to grow. But they lose 90% of the water from the plant through the same holes. Due to the increase in CO_2 levels over the last century, the plants can now obtain the the gas more easily and the stomata have reduced in size. As a result of the smaller holes, plants do not lose so much water and they can thrive in drier regions.

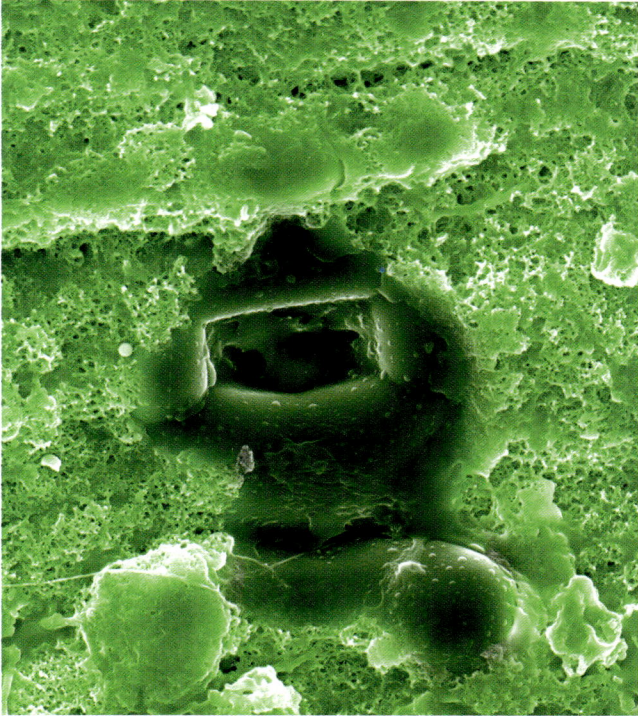

2. Increased absorption of CO2 by the Oceans

The Oceans have absorbed 40% of the Carbon Dioxide caused by the human race and they continue to extract very significant amounts that we currently produce.

(Quay et al. Science 256(5053) pp74-79)

A recent study at Exeter University, UK, has concluded that the Oceans are extracting more Carbon Dioxide from the atmosphere than previously believed.

(Shutler and Watson, Exeter Uni., September 2023

Over the 27 year period, from 1992 to 2018, they state that the sea has absorbed 67 billion tonnes of CO_2 instead of the 43 billion tons originally thought.

They identified that the surface waters which absorbed the CO_2 also cooled which resulted in them sinking allowing fresh surface waters to absorb the gas in turn. Thus there is an aqueous conveyor current which continually pulls CO_2 down to the ocean depths where it remains trapped for hundreds of years.

This increased Carbon Dioxide in the seas is causing increased marine plant growth and it also becomes a component of the shells of sea creatures which, when they die, fall to the ocean depths again causing CO_2 to be removed from the atmosphere.

This Exeter University report also points out that the rate of CO_2 capture is increasing in response to the atmospheric increase.

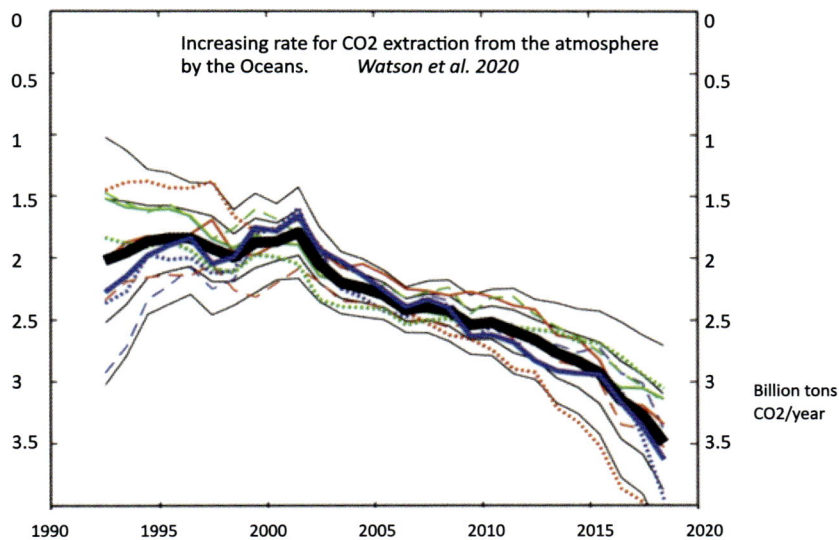

Increasing rate for CO2 extraction from the atmosphere by the Oceans. *Watson et al. 2020*

Billion tons CO2/year

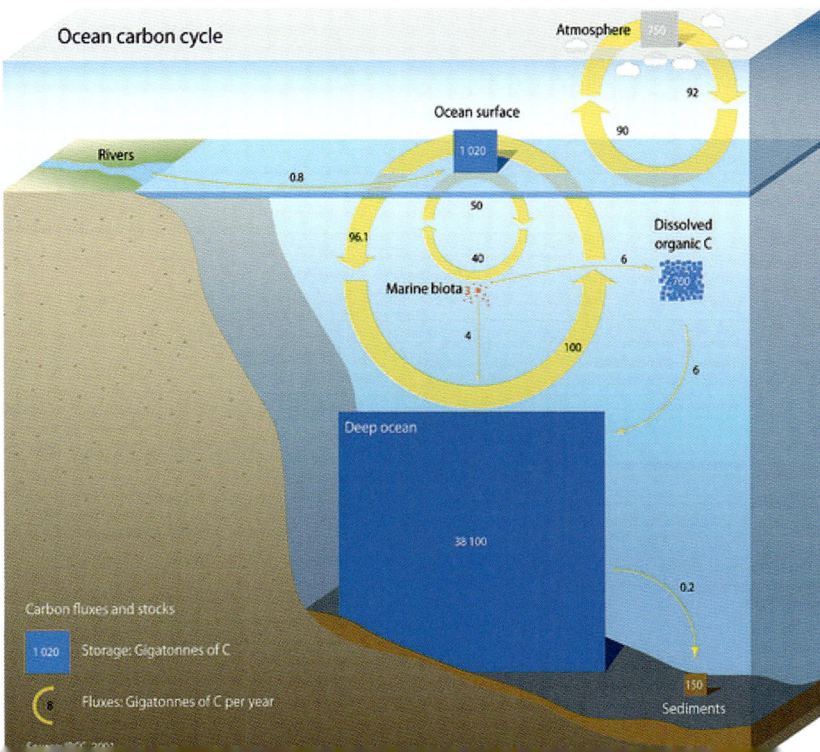

Ocean carbon cycle

Atmosphere

Ocean surface

Rivers

Dissolved organic C

Marine biota

Deep ocean

Sediments

Carbon fluxes and stocks

Storage: Gigatonnes of C

Fluxes: Gigatonnes of C per year

Carbon Dioxide throughout the history of Earth.

In this section we will see how the Carbon Dioxide level and the temperature of the Earth has changed over the time since the Earth was formed.

About 5 billion years ago, the sun, planets and other bodies associated with our solar system began to form from from clouds of dust and gas present in space. The conditions during that era made life on Earth impossible due to extremes of temperature from the new sun, and

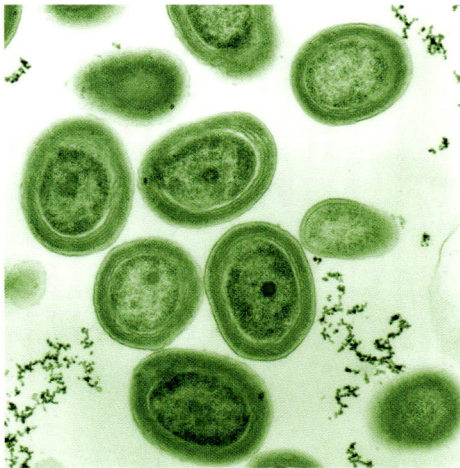

Modern day cyanobacteria.

Credit: Luke Thompson from Chisholm Lab and Nikki Watson from Whitehead, MIT

destructive gases and extensive volcanic action. It was too hot for liquid water to exist, but gradually the Earth cooled down and the water vapour in the atmosphere condensed into rain storms that lasted for many millions of years resulting in the formation of oceans.

By 3.5 billion years ago, the first living forms had formed in the sea from the mix of elements and chemicals that resulted from the torrential storms. These microscopic life forms were an early kind of bacteria.

These cyanobacteria originated at a time when the atmosphere was comprised of a mixture of Methane CH_3, Ammonia NH_3, Carbon Dioxide and other inert gases such as Argon. We are not sure of the relative concentrations of these gases in the atmosphere but we do know that there was less than 1% Oxygen.

Various types of microscopic cyanobacteria evolved and lived in the seas with no known life forms on dry land. They fed on the chemicals in water and used sunlight for energy and as a byproduct they released Oxygen. This caused the oxygen content of the atmosphere to slowly increase so that by around 1 billion years ago there was sufficient Oxygen to enable larger organisms such as plants to thrive and diversify.

As the Oxygen concentration continued to increase, plants began to appear on land and along with the all the life forms in the sea, they thrived on energy from the sun and Carbon Dioxide. As they did so, the Earth was slowly cooling down so that there was a long ice age 720 to 635 million years ago.

We pick up the story at 500 million years ago when the Earth had warmed up and the atmosphere had about 10% Oxygen. This caused a rapid 'explosion' of animal life and the resulting fossils along with geological data enable us to determine the Carbon Dioxide level so that it can be tracked along with the conditions on Earth from 500 million years ago to the present age. We will briefly note what was happening to the CO_2 levels in each of the geological time periods.

The next two pages show the temperature of the Earth and the CO_2 level throughout this 500 million year period. It has been compiled using data from:
Berner & Kothavala, American Journal of Science, vol 301, 2001, pp182-204.

Temperatures from CR Scotese Paleomap, 2016

Note that 420ppm is used as the value for CO_2 level today.

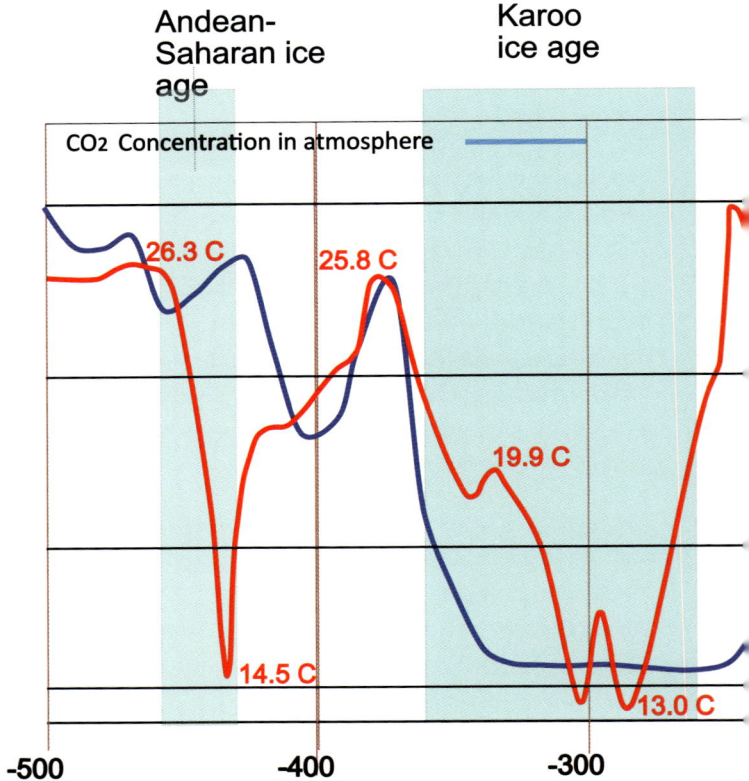

Temperature of the Earth, an[d]

Andean-Saharan ice age

Karoo ice age

CO2 Concentration in atmosphere

26.3 C

25.8 C

19.9 C

14.5 C

13.0 C

-500   -400   -300

This chart shows the temperature of the Earth, as a red line, for the
last 500 million years. It also shows the CO_2 concentration in blue.
shows that the CO_2 concentration has never been as low as it is
now. Note that the small increase in CO_2 which has occurred
during the last 100 years is too small to be shown at this scale.

If we look at the series of ice ages known as the Karoo ice age, the
CO_2 level was 2 times what it is today even though the ice caps
extended over a much greater area than they do now.

Quaternary
ice age

Temperature of the Earth ———

CO2 15x today

24,3 C

CO2 10x today

19.1 C

CO2 5x today

16.3 C

15.0 C

CO2 today

-100   0

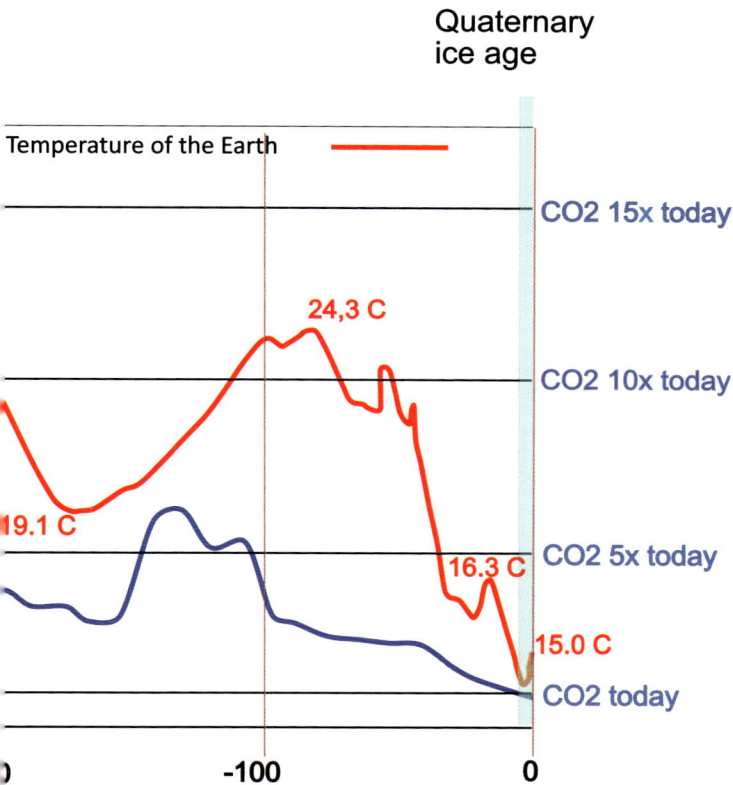

When the Karoo ice age ended and the Earth began to warm up,
the CO_2 level was actually falling which can be expected since
plants were evolving and spreading during this period : hence
they absorbed the abundant CO_2 .

Looking at the chart, it is obvious that there is no close correlation
between the CO_2 level and the temperature of the Earth. There
are so many other factors that determine the temperature and we
can see here that CO_2 is not the dominant cause.

Ordovician to Silurian, 485 - 419 million years ago

When the Ordovician period started the CO_2 concentration was about 6300 ppm: 15 times higher than we have at present. At the same time the average Earth temperature was much hotter than we experience: about 30C compared to the average of 15C we have in today. There was also about 10% Oxygen and this triggered an increase in the number and diversity of living creatures in the seas, lakes and rivers.

Ordovician-Silurian boundary on Hovedøya, Norway,
Petter Bøckman - Own work

The land masses were not colonised during this period which saw the continents shifting across the surface of the planet accompanied by a great number of active volcanoes.

During the Ordovician period, the CO_2 level fell to 5000 ppm. Despite this being over 10 times higher than we experience today, the Earth had cooled down so much that there was an ice age lasting about 20 million years . This caused a mass extinction of a great number of the animal species that had evolved during this period.

This great extinction resulted in habitats which were ideal for a new stage of evolution. So at the start of the Silurian period, the first vertebrate fish with jaws are found enabling them to swim quickly, catch prey and eat plants. It was during this period from 444 - 419 million years ago that plants and animals started to live first at the shores of seas and lakes, and then developed to live on land.

Climate changes during the from 485 to 419 million years ago

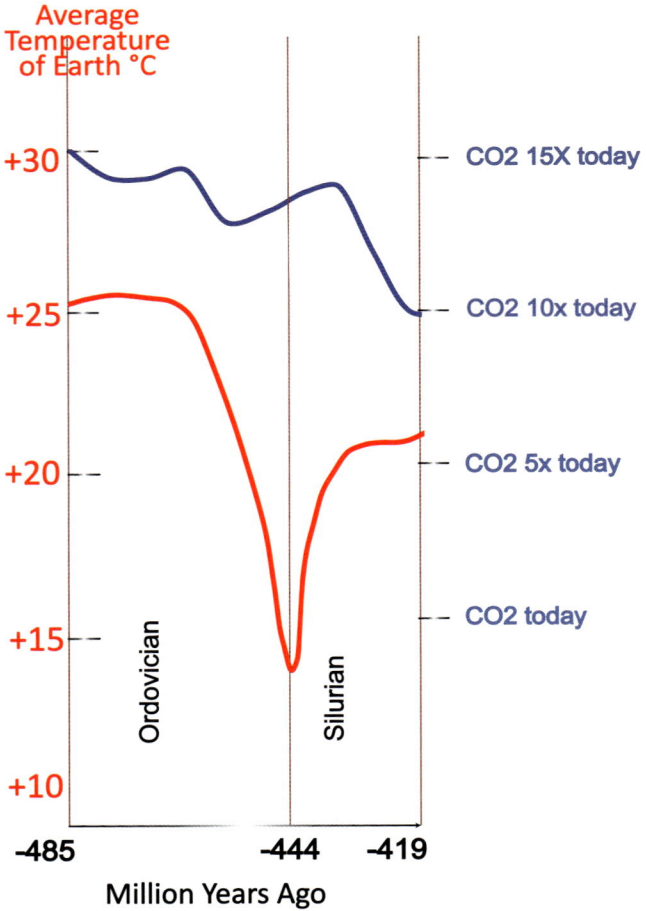

Ordovician 485 - 444 and Silurian 444 - 419 million years ago

During the Ordovician period, the CO_2 level fell from 6300ppm to 5000 ppm. You can see from the chart that the temperature fell for the second half of the Ordovician period causing a long ice age. Not also that the CO_2 level was 13 time higher than it is today at the same time as the ice age.

During the Silurian period the CO_2 level fell back from 5000ppm to 4200ppm.

Also, there was major reduction in volcanic activity which reduced the atmospheric dust. The clear skies allowed the sun to warm the Earth's surface and the ice age came to an end with the Earth's average temperature rising back to 22C (72F) during the period when the CO_2 level was falling.

Climate changes from 419 to 299 million years ago.

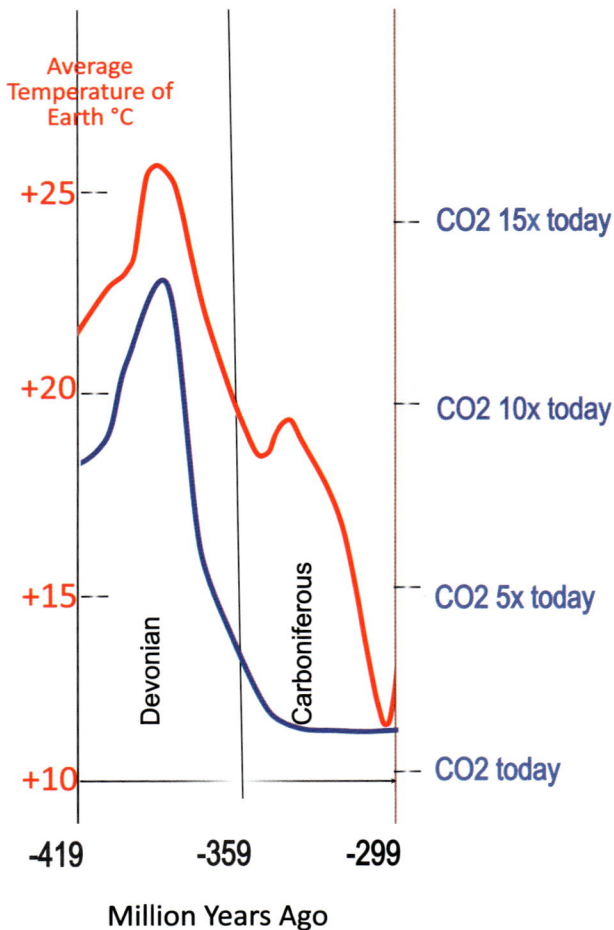

Devonian 419 - 359 and Carboniferous 359 - 299 Million Years Ago

The Devonian is the period when new groups of fish appeared some of which were the ancestors of four legged land animals. On land, plants evolved into new groups and as the habitats diversified, so did the land creatures. The CO2 level decreased from 4200ppm to 900 ppm during this period as plants thrived in this environment.

The temperature of the Earth fell a lot towards the end of the Carboniferous which resulted in extensive ice caps at the North and South poles even though the CO2 level was more than 2x higher than today Carboniferous period is considered to be the time when life on land flourished more than ever on Earth. The climate was temperate and the land masses had converged to form the main continents. Plants, insects , anthropods and amphibians abounded in the warm swamps. The first small reptiles also appeared.

Plant life flourished everywhere which would eventually lead to the production of top soils which are crucial for food production.

Climate changes from 299 to 201 million years ago.

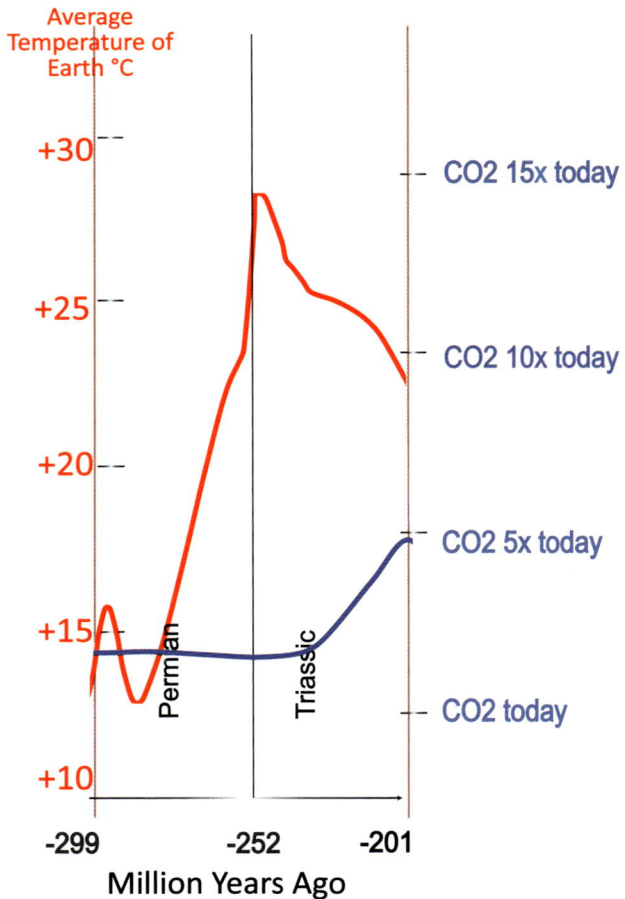

Permian 299 - 252 and Triassic 252 - 201 Million Years Ago

The Permian Period started with an ice age which ended as the average temperature rose from 13C to 28C whilst the CO_2 went down from 900ppm to 800ppm. The land mass, Pangaea, became dryer resulting in a large desert area in its central region. This caused the amphibian population to decline, leading to the evolution of animals that could survive in drier conditions.

Some were as large as 4.5m and 250kg.

D. Giganhomogenes, Nobu Tamura

In the Triassic period, Pangaea remained warm, and experienced summer and winter seasons. Rainfall gradually increased around the coastal regions. This was the age when the reptiles became widespread.

The CO_2 levels increased from 800 to 1900ppm and the average temperature went down from 28C to 23C over this period.

Climate changes from 201 to 66 million years ago.

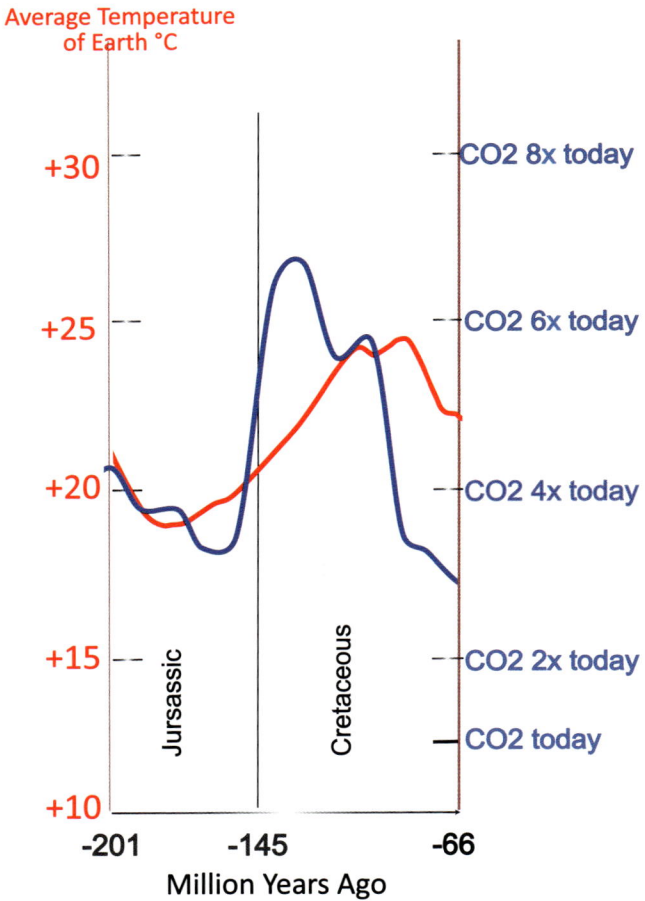

Jurassic 201 - 145 and Cretaceous 145 - 66 Million Years Ago

In the Jurassic period, the large super continent, Pangaea, gradually split into two which caused the patterns of ocean currents and wind directions to change resulting in increased diversity of habitat and of animal species. In water, this is the age of the reptiles and on land, the age of the dinosaurs.

The chart shows how the CO_2 level increased from 1900ppm to 2800ppm during this period and the temperature of the Earth was more or less the same at the beginning and end of the Jurassic.

The Cretaceous period was also dominated by dinosaurs on the land and reptiles in the oceans and rivers. The two land masses called Laurasia and Gondwana continued to drift apart to eventually become the Asian/African and North/South American continents that we know today. There was a mass extinction towards the end of this period which killed all non-bird dinosaurs.

We note from the chart that the temperature of Earth rose and then fell during this period, and see that the CO_2 declined from a peak at 2800ppm to 1200ppm.

Climate changes from 66 to 3 million years ago.

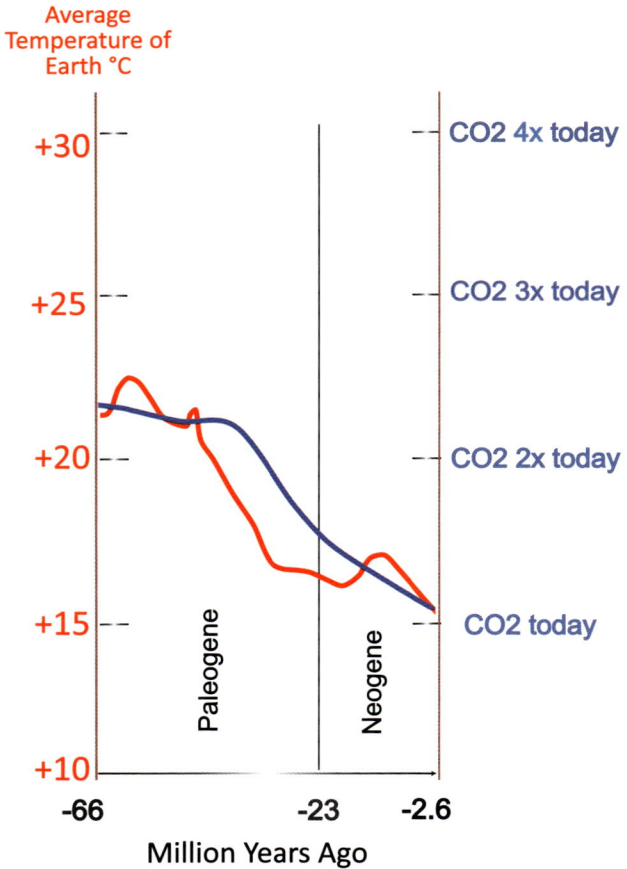

Average Temperature of Earth °C

+30 ---   -- CO2 4x today

+25 ---   -- CO2 3x today

+20 ---   -- CO2 2x today

+15 ---   -- CO2 today

+10

Paleogene   Neogene

-66   -23   -2.6

Million Years Ago

Paleogene 66 - 23 and Neogene 23 - 2.58 Million Years Ago

The mass extinction which occurred at the end of the Cretaceous period resulted in a world where the land and rivers recovered so that the vegetation which had provided food for the dinosaurs was now available for newly evolving animals. These were the mammals and birds. The large islands such as Australia separated from the main continents during this period which led to distinct differences in the animals which evolved on the different land masses.

There were some small rat-like mammals before this period, but as time passed by the variety and size increased due to the luscious vegetation that flourished in the increasingly temperate climate.

We note that the average temperature of the Earth was cooling down throughout the 63 million years of these two periods and at the same time, the CO_2 in the atmosphere was also constantly declining.

Quaternary Ice Age 2.6 million years ago to the present day

In the year this book was written, 2024, we are currently living in the geological period known as the Quaternary Ice Age. During this period there have been extensive ice caps which have advanced and receded several times whilst the CO_2 concentration in the Earth's atmosphere ranged from 290 to 420ppm.

Despite what is being claimed by many in recent times we note that the average CO_2 concentration at the stage of the quaternary period we are living in is the lowest that it has ever been throughout the entire 500 million year evolutionary period of the Earth.

The maximum extent of the Northern ice cap during the Quaternary period By Ittiz - Own work, CC BY-SA 3.0, https://commons.wikimedia.org/w/index.php?

Our current geologic period (Quaternary) has the lowest average CO$_2$ levels in the history of the Earth.

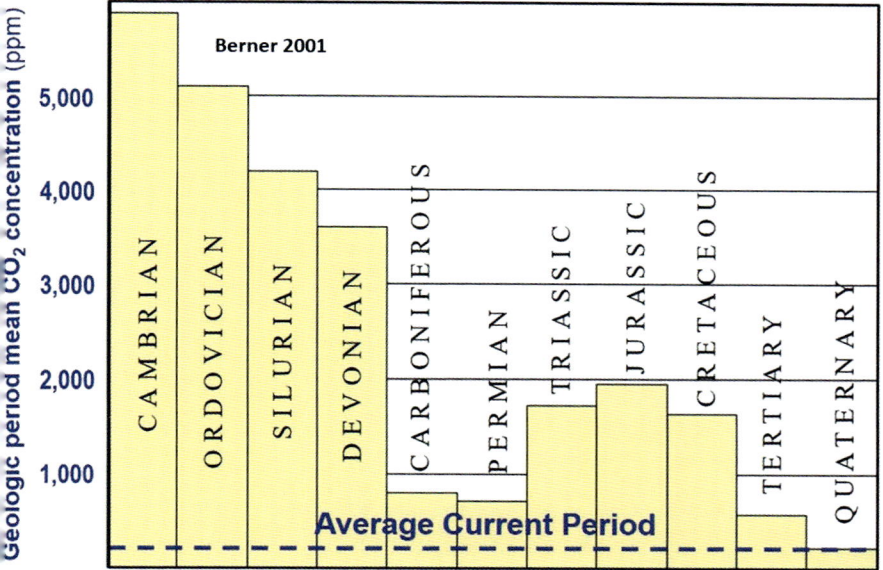

Berner 2001

Geologic period mean CO$_2$ concentration (ppm)

5,000 — 4,000 — 3,000 — 2,000 — 1,000

CAMBRIAN · ORDOVICIAN · SILURIAN · DEVONIAN · CARBONIFEROUS · PERMIAN · TRIASSIC · JURASSIC · CRETACEOUS · TERTIARY · QUATERNARY

Average Current Period

This prolonged reduction in the carbon dioxide concentration is a major factor in causing significant changes to the habitat on Earth and the animals that live on it.

The several periods of extended ice caps greatly influenced the evolution and survival of plants and animals with some animals becoming extinct due to the climate changes, such as woolly mammoths and sabre-tooth tigers amongst them.

During this the Quaternary epoch, deserts began to extend across great areas of the continents. One contributor to this was reduced rainfall in the interior regions. But another was the fact that below 400 ppm CO_2, many plants and trees struggle to grow. In the periods when the Earth was covered with lush vegetation, the CO_2 concentration was greater than 1000 ppm.

The desertification during the last 2.6 million years is now so extensive, that 41% of the land surface is now considered to be arid and supports very little life. It is true that another cause of the desertification is over grazing and poor farming techniques by human beings. However, this only applies to the most recent 10,000 years. The growth of deserts and loss of topsoil has been occurring for the last 2 million years.

Plants cannot grow without enough CO_2, and animals cannot live without plants. This applies to sea life just as much as on land. The low concentration that is currently found in our atmosphere is effectively slowly and gradually causing deserts to spread, which in turn is causing many animal species to become extinct.

We are, in effect, living on a planet that is very slowly dying.

There are several projects being run to try and revitalise some desert regions. But these are not proving to be very beneficial. Introducing irrigation and fertiliser does help, but in order to obtain useful yields for farm crops, then CO_2 is injected into greenhouses to raise its concentration to 1000ppm. - a level that existed in the atmosphere when plants flourished and deserts did not exist. Such a technique cannot be used to bring life back to the extensive deserts that currently exist.

Desertification of the land surface where plants will begin to return if CO₂ levels increase.

Photo Courtesy of the World Economic Furum.

NORTH AMERICA

Sahara Désert

Desert of north America

SOUTH AMERICA

Namib de[...]

Atakama desert

Kalahari désert

Arid regions and deserts cover 41% of the Earth's surface.

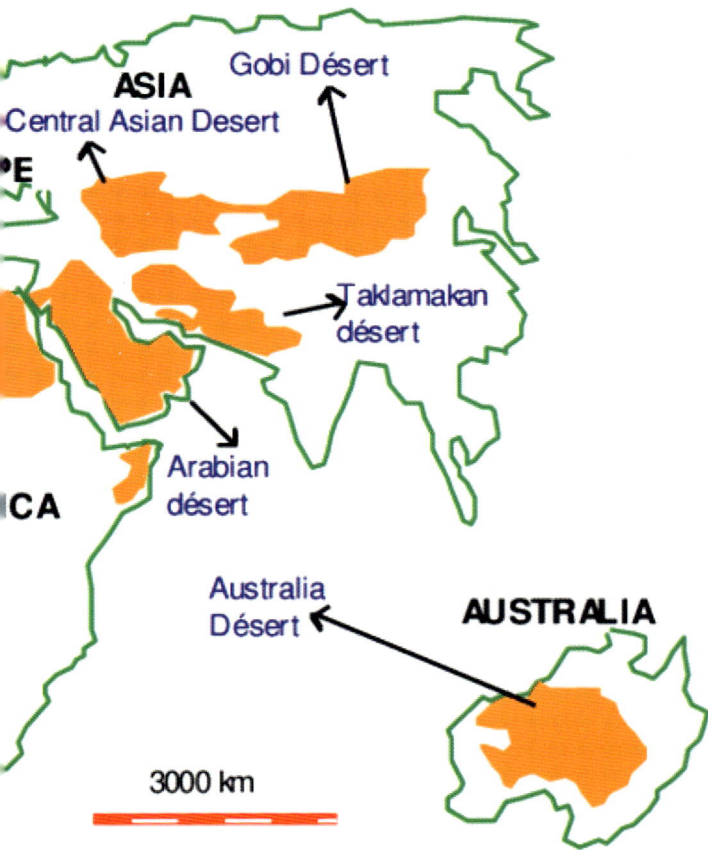

The increase in CO2 levels from 280 to 420 ppm has led to an increase in plant cover of 11%

The Period of Homo Sapiens.

It is generally accepted that the earliest evidence for human beings on Earth dates from 315,000 years ago. We will next look at what happened to the climate during the most recent 450,000 years. The top chart shows the temperature change and the bottom chart shows actual CO_2 concentration over the same period.

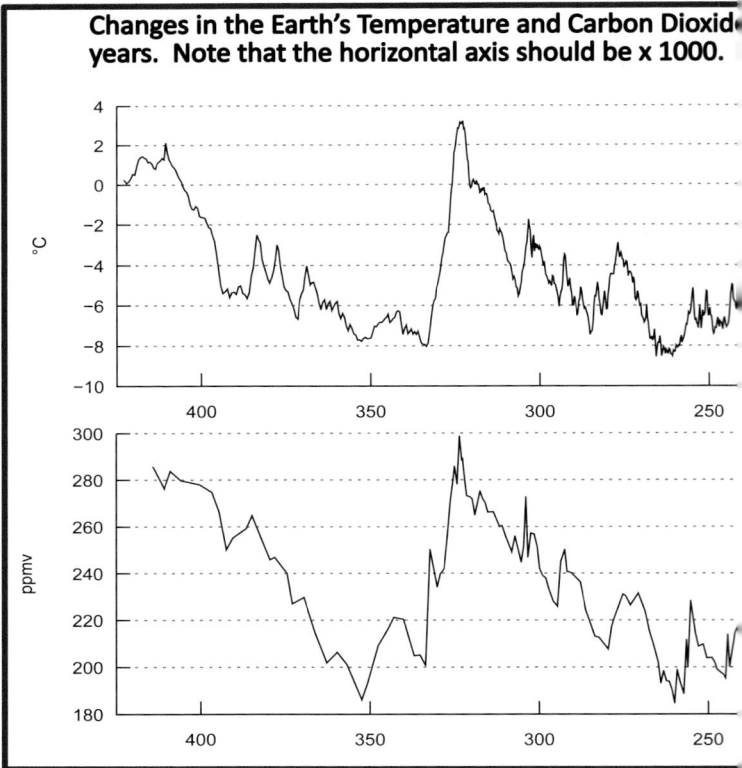

Changes in the Earth's Temperature and Carbon Dioxide years. Note that the horizontal axis should be x 1000.

Allen et al, MPRA Paper No. 103862, posted 02 Nov 2020 15:49 UTC

You can see that about 450,000 years ago the Earth was warmer and this gradually cooled down to cause an ice age.

This happened four times to give four ice ages.

We are living at the right hand side of the chart where we are coming out of an ice age and the Earth is warming up. Note that the CO_2 concentration tracks the temperature raising three questions:

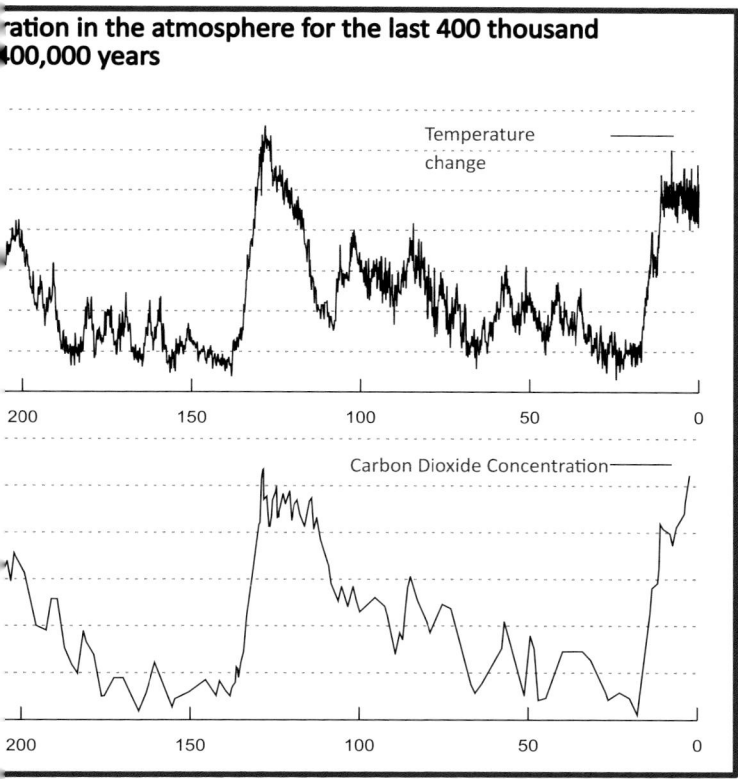

ation in the atmosphere for the last 400 thousand
00,000 years

200	150	100	50	0

Temperature change

200	150	100	50	0

Carbon Dioxide Concentration

1.   Does the increase in temperature cause the increase in CO_2, or does CO_2 cause the temperature rise?

2.   Why does the temperature fall once it reaches a certain peak?

3.   Is the correlation of CO_2 and temperature simply coincidental and other factors are causing this effect?

The chart on the previous pages shows that there have been four periods when th+e Earth was significantly colder than it is today. During those cold periods, the ice at the North and South poles reached further than we see today and they are referred to as ice ages. We are still experiencing the most recent ice age since we still have ice caps throughout the year. We are, however coming out of this ice age and the Earth is warming up as it has done after the previous ice ages.

But just because the CO_2 concentration is rising at the same time as the average temperature is rising, we cannot automatically assume that the the increase in CO_2 is causing the increasing temperature.

When the Earth is colder and much of it is frozen, then the growth of plants is restricted and even prevented at some latitudes. The reduced vegetation and animal life is therefore not giving off so much CO_2 is it would do when there is a warmer tropical climate, resulting in its reduced concentration. When the Earth warms up resulting in increasing plant and animal life, then we would expect the

CO_2 level to increase. In other words, the increase in CO_2 is caused by the warming Earth, and in this scenario, the CO_2 is not itself causing the temperature to increase.

We can see from the temperature/CO_2 data for the past 500 million years that there is no correlation between these two parameters. We should, therefore, not take the simplistic view that CO_2 is the main culprit for the current global warming. And if it isn't, then what is?

Looking at the chart on the previous page there are two features that indicate what is happening.

We can see that CO_2 and temperature rise relatively quickly, reach a peak, and then fall back again at a slower rate. Why does the temperature not keep rising? What causes it to fall back. There is likely to be another factor or factors that are causing this to happen. In this case, there is data which shows that when the temperature rises, then the dust in the atmosphere also increases. This can be due to winds, clouds, volcanoes and other natural phenomena. This dust, in the form of aerosols, reaches the upper atmosphere and reflects some of the sunlight away from our planet. As the temperature increases, so does the dust concentration until it reaches a point that it causes the Earth to cool down despite the existence of greenhouse gases.

Cooling caused by dust.

The Earth's atmosphere contains a lot of dust, often referred to as aerosols since the particles are small enough to remain airborne for long periods of time.

Dust clouds in the upper atmosphere due to a storm in the Sahara desert.

NASA space station.

The global dust mass has increased by 55% since pre-industrial times originating mainly from the deserts of North Africa and Asia, rising up as high as 6 to 20km, (4 - 12 miles). Researchers state that the mass of the fine dust ranges between 20 - 40 million tons.

These fine particles capture some of the light from the sun and warm up the rarefied atmosphere. This then re-radiates the heat , some of which goes back into space, with the result that less heat reaches the surface of the planet. In this way, the dust produces a cooling effect which counteracts global warming. Current estimate is that the reduction in heat at the Earth's surface is -0.2 Watts per square metre.

As the planet gets hotter, then the dust concentration increases, and the heat radiated back to space increases exponentially serving to produce ever increasing cooling.

This effect has not been sufficiently incorporated into current global warming models.

Jasper Koh et al, Nature Reviews Earth & Environment April 2023

Cooling caused by volcanic eruptions.

Active volcanoes inject ash and gases into the stratosphere. The ash, seen as clouds of dust sinks back to the ground after a few days and has little effect on the Earth's temperature. The most significant component is sulphur dioxide which is condenses rapidly to form fine aerosols of tiny droplets of sulphuric acid. The aerosols remain suspended in the upper atmosphere for prolonged periods where they act as a reflector causing some of the radiation from the Sun to be reflected back into space.

During the last century there have been several volcanic eruptions which caused a decease in the average temperature of our planet by up to half a degree.

The climatic eruption of Mount Pinatubo in the Philippines on June 12 - 15th 1991 released a cloud of 20,000,000 tons of sulphur dioxide into the stratosphere at a height of more than 35km, (20 miles). This resulted in a worldwide dispersal of aerosol which lowered the average temperature of the Earth by 0.8 deg C (1.3 deg F) and persisted for a period of three years.

Temperature changes caused by planetary motion.

Another feature of the chart on pages 68 - 69 is that although it gives data spanning 450,000 years, the rise and fall of the Earths temperature has a repeatable periodicity: it looks like something caused by a mechanism, such as a mechanical clock. And when we look at the way the Earth rotates about it's own axis, and around the sun, this is basically a mechanical performance which directly controls the amount of energy arriving at different latitudes and at different times of the year. There are some very long term cyclical changes that do result in warming and cooling periods. The effects were first noted by the Serbian geophysicist and astronomer Milutin Milankovitch.

He determined that variations in eccentricity, axial tilt, and precession of the Earths rotation together with long term changes in the shape of its orbit around the sun combined to result in cyclical variations in the year to year latitudinal distribution of solar radiation at the Earth's surface, and that this strongly influenced the Earth's climatic patterns.

Milankovitch Cycles

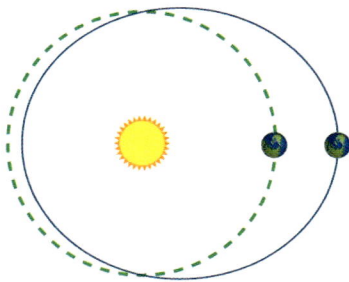

Eccentricity   Obliquity (Tilt)   Orbital Precession

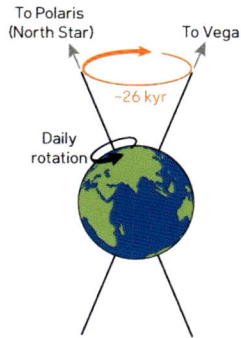

These variations in the Earth's orbit have the repeat periods over the following time frames:

Orbital Precession where the axis of its rotation itself rotates every 25,771 years. This 'wobble' is not perfectly smooth since there are also gravitational effects from the positions of Saturn and Jupiter. Nevertheless, the current position of the axis is tending to make the Northern winters less cold.

Obliquity, i.e. the angle of tilt of the Earth ranges between 22.5 and 24.5 degrees with a periodicity of 41,000 years. It is currently, slowly moving towards the 22.5 angle which is gradually making the winters warmer in the Northern Hemisphere.

Eccentricity, refers to the shape of the Earth's orbit around the sun. This changes from being almost circular to its most extreme eccentricity and back again over a period of about 100,000 years, due to gravitational attractions from Jupiter and Saturn. The eccentricity is relatively small, but does result in less solar radiation hitting the Earth during the most eccentric phase. We are currently approaching the circular phase of this effect.

The Eccentricity effect first reported by Milankovitch fits very closely to the ice ages and warmer periods shown in the chart on page 40/41.

The actual effect of the four Milankovitch cycles is difficult to analyse bearing in mind that they are taking place simultaneously with different periodicities. They definitely do have an influence on global warming and cooling and much research and discussion amongst scientists takes place on this subject.

But it seems that these effects are currently being disregarded, maybe because they are complex. But it is clear that the recent ice ages and warming periods do match the Milankovitch predictions due to the remarkable periodicity.

What is Net Zero?

Net Zero is a concept promoted by the United Nations and subsequently agreed by 140 of the largest nations on earth. It is based entirely on the premise that the current warming of the Earth during the last 100 years is primarily caused by the increased CO_2 emissions due to the lifestyle of the increasing human population of the Earth. The claim is that in order to preserve a liveable planet, global temperature increase needs to be limited to 1.5°C above pre-industrial levels. The UN points out that the Earth is already about 1.1°C warmer than it was in the late 1800s and that emissions continue to rise. In an attempt to achieve this, an international agreement was signed in Paris which stated that to keep global warming to no more than 1.5°C, that emissions need to be reduced by 45% by 2030, and reach net zero by 2050.

The assumption is made that the 'greenhouse' gases, CO_2 and methane are directly responsible for global warming.

The UN does not mention that the most dominant greenhouse gas by far is water vapour.

The scientific evidence presented in this book shows that that we cannot expect the temperature of the planet to be controlled by the CO_2 we produce. In the 500 million years that we have reliable data, sometimes the planet gets warmer, much warmer, when the CO_2 level goes down - and sometimes it gets cooler when CO_2 increases. There is no correlation between these two parameters.

By using 'the late 1800s as the reference point the UN is exaggerating the current warming period, since at that time the Earth's climate is referred to as 'the little ice-age'.

The means by which the UN proposes to reduce the global warming is to reduce the use of fossil fuels - coal and oil and replace these by the use of solar energy and wind energy referred to as 'renewable energy'. The cost of doing this is enormous and will affect the life of everyone on Earth.

Politicians from the countries that have signed up to net zero are now dedicated to implementing the agreement regardless of the cost. This is outlined in a report by the consultancy firm McKinsey who state, in a report titled 'The Net-Zero Transition: What it Would Cost,', that global spending on changing to solar and wind energy along with the extensive new electricity transmission lines will need to increase by $3.5 trillion every year until 2050.

As a result, huge sums of money are being given to many

multi-million dollar schemes and research projects which has created proponents in industries and universities whose income now depends upon this source of funding.

This, in turn, means that for the governments, businesses and university research teams that have signed up to the concept of the climate catastrophe, they no longer question whether or not the reduction in CO_2 emissions they hope to achieve will have any significant effect on the temperature.

The McKinsey report which was written for the World Economic Forum goes on to say that the most noticeable impacts on everyday lives will include rising energy bills, job losses in high-emission industries, changes in what people eat, and increasing outgoings to end our dependence on fossil fuels to heat homes and travel.

Consumers will face the cost of replacing home heating systems and cars that run on fossil fuels, and will have to change their diets to avoid high-emission foods such as meat.

In order to convince the general population of the need to make these major changes to our lives, governments and the media have created a consensus which overwhelmingly supports the view that we must reduce CO_2 emissions.

Now, almost any example of an extreme weather event is claimed to be an example of global warming. And the impression is given that the global warming link with CO_2

emissions is accepted by all the scientific community. There are, however, many experienced and respected scientists whose research shows that CO_2 concentration is not the dominant factor in controlling the Earth's temperature.

In a recent paper, [*Net Zero Averted Temperature Increase, Lindzen, Happer and Wijngaarden, MIT, June 2024*], the authors make an analysis of the result to be expected when the net zero plan, to reduce Carbon Dioxide levels to pre-industrial levels, is carried out. They conclude that if the entire world forced net zero CO_2 emissions by the year 2050, the reduction in warming would be only 0.28°C (0.50°F). This conclusion uses figures provided by the Intergovernmental Panel on Climate Change (IPCC). And the result is based on the understanding that the entire population of the planet will participate in implementing the net zero actions.

This may be a surprising conclusion, but it arises because the relationship between CO_2 concentration and temperature is not linear. As CO_2 levels increase, this has a smaller and smaller effect on temperature.

It is naive to believe that the modifications to our lifestyle proposed by the net zero project will have a large enough effect to significantly change the global climate.

Simply looking at the historical temperature changes over the last 450,000 years shows that similar climate changes to those we are currently experiencing have occurred when there was little or no human population and absolutely no industrial use of fossil fuels.

Another result of net zero is that the human population would decline. This is considered a good thing by some. But where human habitation declines the territory would be reclaimed by insects and other wildlife that themselves emit CO_2.

Yes, we are experiencing global warming, but maybe the best we can do is to accept this phase of climate change that the Earth is passing through, and spend the money that is currently being allocated to net zero, on means of living with the slightly higher temperatures.

During the second half of the 20th Century, when the Earth was cooling and there was talk of an ice-age coming, then money was spent on keeping warm. Maybe the money should now be spent on keeping cool such as on better insulation.

In the conclusion to the paper by Lindzen et al, they state '….. there appears to be no credible scenario where driving … emissions of CO_2 to zero by the year 2050 would avert a temperature increase of more than a few hundredths of a degree centigrade. The immense costs and sacrifices involved would lead to a reduction in warming approximately equal to the measurement uncertainty. It would be hard to find a better example of a policy of all pain and no gain.'

Conclusions

1. Looking back over the 500 million years during which life was evolving on Earth, we see that the Carbon Dioxide concentration in the atmosphere varied between 300 and 1200 parts per million. We are currently at 420 ppm which is near the lowest concentration throughout that 500 million year period.

2. There have been major ice-ages lasting millions of years when the CO_2 level was up to three times greater than it is now. This shows that high CO_2 concentrations did not cause global warming during those periods. We conclude that there are other, more dominant effects that control the Earth's temperature, and although CO_2 is a greenhouse gas, it did not automatically cause an increase in temperature.

3. A major factor causing a gradual rise in temperature is that the changing orbit of the Earth around the Sun is becoming less elliptical and approaching that of a circle which means it is receiving more energy. This effect has coincided with the more recent ice age/ warming cycles. We cannot do anything about this

4. There are multiple other events that influence our temperature, some predictable such as the Sun spot 11 year cycle and El Niño (the warming of the Pacific Ocean) which occurs every 2 to 7 years, and some unpredictable events such as volcanic eruptions, and prolonged changes in the wind direction. These are also outside our control.

5. The relatively low CO_2 concentration we are currently experiencing is part of the reason why there are extensive deserts on Earth. The more recent increase

from 300 ppm to 420 ppm has been accompanied by the plant cover of the world increasing, and farming crops have higher yields. This has been responsible for improving the health of the entire population particularly in poorer countries. Carbon Dioxide is essential for all life on Earth.

6. The climate is always changing; it has been doing so for all of the past and will continue to do so.

7. There is no guarantee that the current major emphasis on limiting CO_2 emissions from human activities will have any effect on limiting global warming. In fact, the evidence highlighted in this book suggests that the human endeavour will not only be very expensive, but will achieve nothing.

On 12 December 2015 representatives from 196 countries attended the UN Climate Change Conference (COP21) in Paris, France, They signed a legally binding international treaty on climate change which entered into force on 4 November 2016. The treaty states that any future temperature rise on Earth should be limited to 2°C and preferably the limit of the increase should only be 1.5 °C (2.7 °F).

It also says that to achieve this temperature goal, greenhouse gas emissions should be reduced as soon as, and by as much as, possible. They should even reach net zero by the middle of the 21st century. To stay below 1.5 °C of global warming, emissions need to be cut by roughly 50% by 2030.

This will involve vast payments to invest in means of reducing such emissions, it will reduce the quality of life for many people, will increase poverty levels, and reduce life expectancy - and will most likely have zero effect on the temperature of our planet.